NEXTGEN LEARNERS PRESENTS

Think Like a Brain!™ Collection

Amy the Amygdala & The Overactive Alarm

Why Big Feelings Happen and How to Handle Them

BOOK 8

Think Like a Brain!™
Collection Overview

Think Like a Brain!™ is a wildly engaging, silly-smart children's book collection that invites kids to explore how their brains work from the inside out. Through playful characters, imaginative settings, and rhythm-rich storytelling, this collection teaches neuroscience, social-emotional learning (SEL), and self-understanding in ways children absorb without even realizing they are learning.

Instead of telling children how to behave or what to feel, Think Like a Brain!™ shows them why their brains respond the way they do. Each story transforms complex brain science into friendly characters and relatable moments, helping kids feel capable, curious, and empowered rather than confused or ashamed.

Designed for children first and embraced by parents and educators as preventative education, this collection supports:

- Emotional regulation and self-awareness
- Executive functioning skills
- Neurodivergent understanding and inclusion
- Brain-body connection and growth
- Compassion for self and others

With humor, heart, and factual accuracy at its core, Think Like a Brain!™ helps children build the skills many adults wish they had learned sooner, one playful page at a time.

One of the collection's most powerful features is how the characters become real-life reminders for children long after the story ends. Once kids learn who the characters are and what jobs they do, they can begin to recognize those same processes in themselves. A big feeling might become, "It looks like Amy is trying to tell you something," or a moment of pause might sound like, "What does your Fred think?" By naming experiences through familiar characters, children learn to notice what is happening inside their brains, reflect on it, and move forward more quickly with understanding rather than frustration.

Series 1:

The Brainy Bunch (Brain Anatomy)

Series Overview: Meet the brain's core parts as a friendly team. Kids learn what each part does and how they work together to help thinking, feeling, moving, and learning.

1. **The Brainy Bunch**
   <u>Focus</u>: Brain overview and teamwork
2. **Brainstem Benny & The Automatic Adventure**
   <u>Focus</u>: Breathing, heartbeat, survival basics
3. **Bella Cerebellum's Balance Ballet**
   <u>Focus</u>: Balance, coordination, practice
4. **Olly Occipital & The Lookout Tower**
   <u>Focus</u>: Vision, visual processing
5. **Tammy Temporal & The Soundy Sound Storm**
   <u>Focus</u>: Hearing, language, sound processing
6. **Perry the Parietal & The Body Map Bash**
   <u>Focus</u>: Touch, space, body awareness
7. **Hippocampus Hal & The Memory Maze**
   <u>Focus</u>: Memory, learning, recall
8. **Amy the Amygdala & The Overactive Alarm**
   <u>Focus</u>: Emotional alarm system
9. **Luna Lefty & Roger Righty's Mixed Up Masterpiece**
   <u>Focus</u>: Left/right hemispheres, teamwork
10. **Callie the Connector & The Brain Bridge**
    <u>Focus</u>: Corpus callosum, brain-to-brain communication

Series 2:
Cortex Castle (Frontal Lobe Executive Functions)

Series Overview: A lively castle of executive-function helpers. Kids practice planning, starting, stopping, focusing, flexible thinking, time skills, and self-talk, with Frontal Fred guiding thoughtful choices.

1. **Cortex Castle**
   Focus: Executive function overview
2. **Frontal Fred & The Decision Dilemma**
   Focus: Decision-making, prioritizing
3. **Impulsive Ian & The Button Box**
   Focus: Impulse control, pause
4. **Memory Molly & The Wandering Thoughts**
   Focus: Working memory, holding steps
5. **Focus Frannie & The Great Distraction Chase**
   Focus: Attention, refocusing
6. **Planella & The Checklist Parade**
   Focus: Planning, sequencing
7. **Flexi & The Changey-Changeroo Day**
   Focus: Cognitive flexibility
8. **Self-Talk Sally & The Thoughtful Echoes**
   Focus: Self-talk, emotional regulation
9. **Starter Stacy & Finisher Finn**
   Focus: Task initiation, task completion
10. M**ilo the Minute Keeper & Mae the Moment Seeker**
    Focus: Time awareness, pacing
11. **The Power of Pausing with Penny**
    Focus: Pause skills, regulation

Series 3:
Busy Brain Café (Neurodivergent Affirmations)

Series Overview: A welcoming café where different brain styles are celebrated. These stories affirm neurodivergent experiences while teaching self-understanding, communication, and compassion for others.

1. **Nathan Neuron & The Bouncy Brain**
   Focus: ADHD metaphor, movement, attention
2. **Sara Sprinkle the Sensa & The Super Sensitive Day**
   Focus: Sensory sensitivity, overload tools
3. **Overwhelmed Owen & The Senses Overload**
   Focus: Sensory overload, coping strategies
4. **Interoceptive Ivy & The Inside Signals**
   Focus: Interoception, body signals
5. **Zane's Zigzag Thoughts & The Path to Peace**
   Focus: Anxiety, grounding
6. **The Busy Brain Café**
   Focus: Neurodivergent inclusion, different processing styles
7. **The Listening Light**
   Focus: Nonverbal communication, understanding
8. **The Language of Love**
   Focus: Connection, communication, compassion
9. **Trying Theo & The How-To Hiccup**
   Focus: Persistence, frustration tolerance
10. **Drowsy Drew & The Nighttime Repair**
    Focus: Sleep, brain repair
11. **Burned-Out Bonnie & The Rebuild**
    Focus: Burnout, rest, recovery

Series 4:
Steady Systems Squad (Regulation & Growth)

Series Overview: A squad that explains how brains regulate, reset, rest, and grow stronger over time, including stress waves, sleep crews, sensory filtering, and rebuilding after hard days.

1. **The Steady Systems Squad**
   Focus: Regulation systems overview
2. **Timothy Thalamus & The Traffic Tower**
   Focus: Sensory filtering, attention routing
3. **Danny Dimmer & Wake-Up Wonda**
   Focus: Arousal and alertness
4. **The Start, Stop, and Switch Station**
   Focus: Initiation, inhibition, shifting
5. **The Night Shift & The Sleepy Crew**
   Focus: Sleep systems, restoration
6. **The Stress Wave & The Return to Calm**
   Focus: Stress response, recovery
7. **The Brain Growth Gathering**
   Focus: Brain growth and development
8. **Paul's Practice Path**
   Focus: Neuroplasticity
9. **Daisy's Daydream Network**
   Focus: Daydream network, imagination
10. **Remy's Rebuild Crew**
    Focus: Recovery and healing

NEXTGEN LEARNERS PRESENTS

Think Like a Brain!™ Collection

Amy the Amygdala & The Overactive Alarm

Why Big Feelings Happen and How to Handle Them

BOOK 8

WRITTEN &
ILLUSTRATED BY:

MAISEL MCLAULA

NEXTGEN LEARNERS PRESS™
CINCINNATI, OHIO

This book is published by NextGen Learners Press, an imprint of NextGen Learners LLC.

Maisel McLaula is a pen name used by the author of this work.

First Edition: 2026
Printed in the United States of America
ISBN: 978-1-972860-07-6

This book is intended for educational and informational purposes only. It is not intended to replace professional medical, psychological, or therapeutic advice, diagnosis, or treatment.
Always seek the advice of a qualified professional with any questions regarding a medical or mental health condition.

Artwork and visual elements in this book were created using a combination of original artwork, with enhanced original direction, digital illustration tools, and AI-assisted creative technologies.

A Note for My Readers:

Hi friend,
Your brain is incredible. It thinks, feels, remembers, solves, dances, creates, and so much more!
This series is here to help you understand how your brain works, and to cheer you on as you grow your own thinking superpowers.

Every part of your brain has a job.
Sometimes they get mixed up. That's okay!

When you learn what each part does, you can work with your brain instead of feeling like it's working against you.
So read, giggle, learn, and maybe even draw your own Brainy Bunch when you're done.

I'm cheering you on with every rhyme.

-Maisel McLaula

Think Like a Brain!™

Brain Anatomy Collection

Explore the castle inside your head! Meet each room and its brilliant resident, from Frontal Fred to Benny the Brainstem. Whimsy, wonder, and neurons await!

For Eliana,
A brilliant soul, full of spark, who had big reactions. You showed us how powerful the art of learning how to pause before reacting can be. Your growth and courage inspired this story.

And...

To the beautifully wired, the curiously wired, and the wonderfully neurodivergent minds, may you learn which alarms are real, and which ones can rest.

You are exactly what a great brain looks like.

-Maisel McLaula

In the tallest tower with windows so round,
Lived Amy the Amygdala, alert to all sound.
She kept watch for danger, for fear, fight, or flight,
With eyes wide as saucers both day and night.

She had sirens and whistles, alarms on the wall,
That blasted and boomed when she sensed a close call.

A sneeze, a shadow, a whisper or shout,
Could send her into panic and acting right out!

One morning she spotted a fly near her tea,
She screamed, “CODE RED! It’s attacking me!”

The bells all rang out and the castle turned wild,
Though it was just a bug, tiny, harmless, and mild.

Amy paced in her tower, her alarms still ringing,
Her heart thumped fast, her thoughts kept pinging.

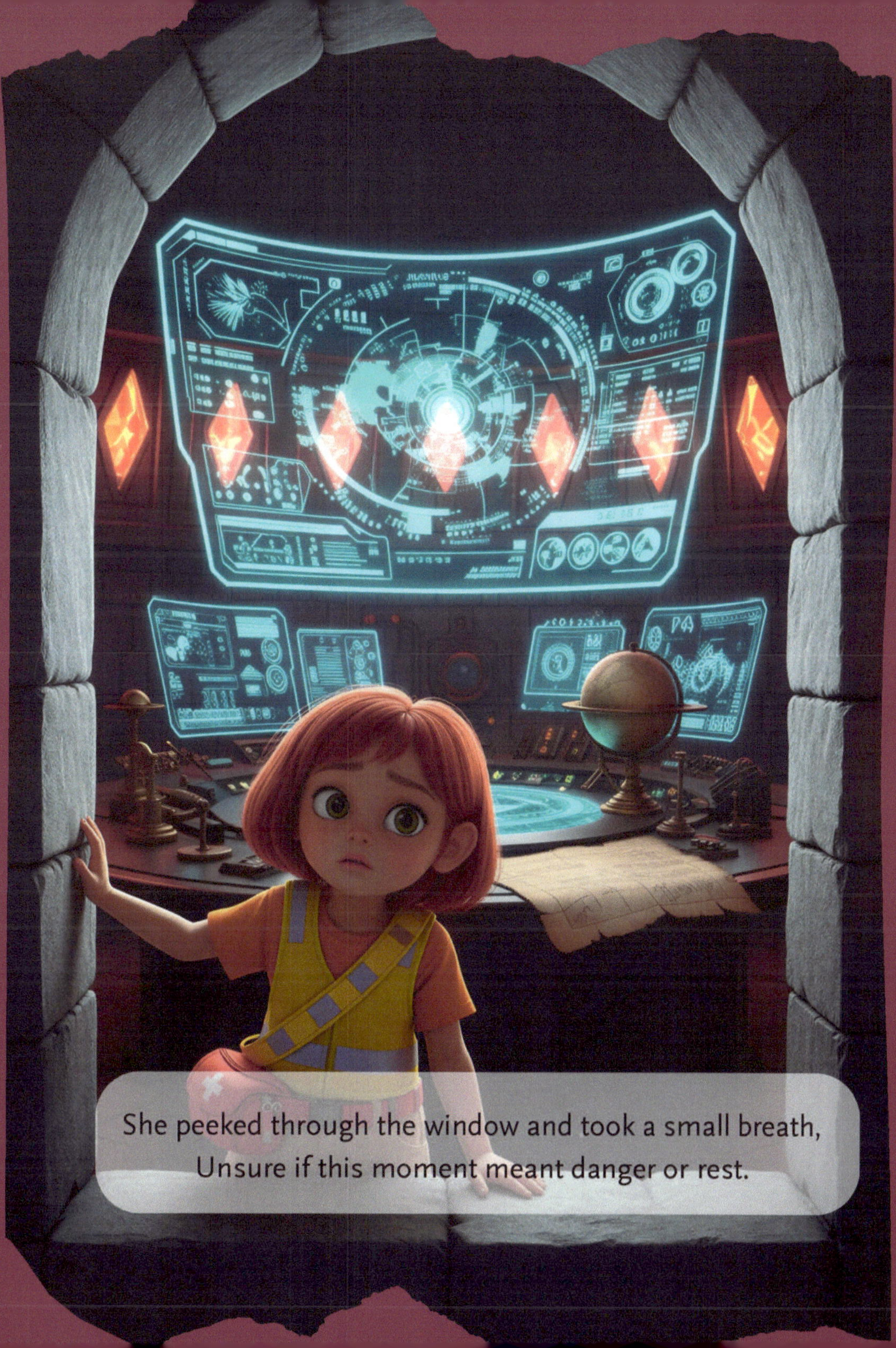

She peeked through the window and took a small breath,
Unsure if this moment meant danger or rest.

Fred dropped his clipboard, Hal dove under a chair,
Tammy covered her ears, there was chaos in the air!

Bella tripped over Perry, who bumped Benny's head.
All because Amy was filled full of dread.

Fred climbed the tower with calm in his tone,
He said, “Amy, your warning light’s overblown.

You're helpful, you're needed, but you go too far,
You sound the alarm for a sneeze or a star!"

Amy blinked twice and looked at the chart,
She saw all the times she'd leapt with her heart.

“Not every small worry is real, big, or bad...
Maybe I panic when I don’t need to be sad.”

Fred gave her a tool: a Pause-and-Check Plan,
With pictures and colors and steps she could scan.

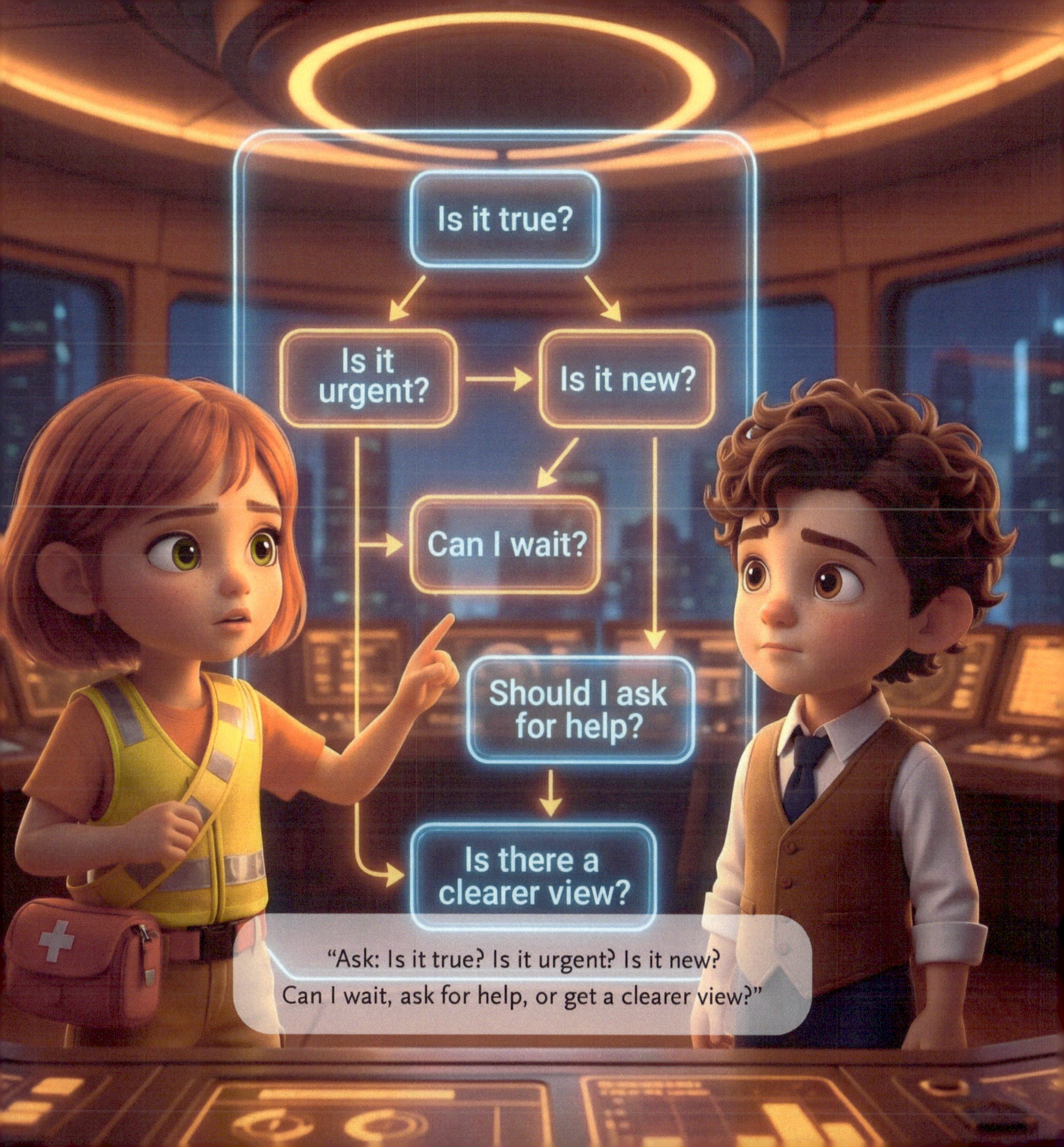
Is it true?
Is it urgent?
Is it new?
Can I wait?
Should I ask for help?
Is there a clearer view?
"Ask: Is it true? Is it urgent? Is it new?
Can I wait, ask for help, or get a clearer view?"

Amy practiced her pause with each rattle and bump,
She labeled her feelings, was it fear, or a grump?

She scribbled her thoughts and took little sips,
And found she could calm her panicky flips.

Next time a balloon popped and echoed like thunder,
She jumped, but she paused, she didn't fall under.

She said, "Just a noise. I'm okay. I can tell."
And the castle stayed peaceful. It all turned out well.

Now Amy still watches with care and with speed,
But she checks first to see if it's really a need.

She keeps her alarms but turns them down low,
So her fear doesn't flare unless it's time to go.

The Brainy Bunch thanked her with banners and signs,
"You're braver than ever, with thoughtful designs!"

She waved from her tower, more calm and less loud,
Still alert, still important, but not panicked, more proud.

So if your own worries start marching in fast,
And your heart feels like thunder that's racing past,

Pause and ask questions, breathe slow and deep,
Not every alarm means a fear you must keep.

Amy's Calm-Down Clues

- Pause before reacting
- Ask: Is this an emergency?
- Label your feeling (fear, surprise, stress)
- Use tools like breathing, drawing, or talking it out

What Is the Amygdala?

Your amygdala helps you sense danger or threat.
It's like your brain's alarm bell.
It's important, but it works best when it checks in
with other brain parts before reacting.

Try It Yourself!

- Draw your own alarm bell or feeling chart
- Create a Pause-and-Check Plan
- Practice calming tools: breathing, counting, doodling
- Roleplay what to do when something startles you

Congratulations!

You met a feeling that felt big and strong,
And learned it doesn't mean something is wrong.
Amy helped you notice, protect, and care,
And showed that feelings are always there.
You listened, you learned, you stayed aware,
That brave little pause showed thoughtful care.
Your brain is growing, steady and true,
Nice work noticing what feelings do.

Thank You to My Readers

Thank you for reading this book.
Whether you're a parent, a teacher,
a therapist, or a brilliant young
brain explorer,
your interest in understanding
how the mind works
means the world.

Sharing these books
means spreading compassion and
curiosity.

You are the reason
these stories come alive.
Thank you for believing in the
power of playful learning.

Stay curious. Stay kind.
Stay brainy!

Amy the Amygdala & The Overactive Alarm

Teacher & Parent Guide

💡 What This Story Teaches:

• Brain part: Amygdala (emotion and safety alarms)

• Core idea: Not every alarm means danger. Pausing helps feelings settle.

• Skills supported: emotional regulation, threat assessment, calming responses

• Big message: Big feelings are signals, not mistakes. We can learn to check the alarm.

Amy shows children that emotions arrive fast, but calm choices arrive when we slow down.

Key Concepts for Grown-Ups

Amy represents the amygdala, a small almond-shaped structure deep in the brain that scans for danger and activates fight, flight, or freeze. It works faster than thinking systems and is designed to protect.

When Amy fires too often or too strongly, children may:

- Have sudden meltdowns
- Become easily startled
- Show anxiety or worry
- React before thinking
- Struggle with transitions
- Appear oppositional when overwhelmed

Important foundations:

- The amygdala reacts before logic.
- Stress, fatigue, hunger, and sensory overload increase alarms.
- Calm input teaches Amy what is truly safe.
- Repeated regulation builds stronger pause pathways.

This story helps children understand that:

- Strong feelings are not bad.
- Their alarm is trying to help.
- They can learn to pause, check, and choose.

🗣️ Discussion Starters

- What made Amy's alarm go off?
- Was it a real danger or a big feeling?
- What helped Amy calm down?
- What does your body feel like when your alarm sounds?
- What helps you pause before reacting?

Encourage children to describe body sensations such as tight muscles, fast breathing, warm cheeks, or shaky hands.

👪 Try This Together:

1. Pause and Breathe

Inhale through the nose for 4 counts. Hold for 4 seconds. Exhale through the mouth for 6. Repeat 3 times.

2. Name the Feeling

Say the emotion out loud: mad, sad, scared, frustrated, excited.

3. Body Reset

Shake arms and legs, stretch tall, then hug knees.

4. Safe Signal

Create a hand signal or word that means "I need a pause."

5. Calm Choice Cards

Write simple options: breathe, drink water, squeeze hands, take space.

🎯 Bonus Challenge

Build Amy's Alarm Station

Invite children to draw or build an alarm control center.

Ask:

- What makes Amy's alarm ring?
- What buttons help turn it down?
- What tools help the body feel safe?

Optional Labels:

- Pause Button
- Breath Dial
- Body Reset Switch
- Calm Meter

Playful Extension:

Real Alarm or False Alarm

Call out scenarios and let children decide:

- Loud thunder
- Spilled milk
- Someone bumps into you
- Fire alarm drill

Practice choosing a calm response for each.

🛠️ Amy's Calm Toolbox:

Use these supports when children experience big emotions:

- Model slow breathing
- Offer water or a snack
- Reduce sensory input
- Validate feelings before problem-solving
- Provide predictable routines
- Teach simple pause scripts
- Practice regulation during calm moments

Helpful phrases:

- "Let's help your body first."
- "That's Amy checking for safety."
- "We can pause together."

Avoid reasoning during peak emotion. Regulate first.

❤️ Final Note for Grown-Ups

Amy reminds us that emotional reactions are protection, not defiance.

Children do not need stronger consequences when alarms sound. They need steadier nervous systems.

Every calm moment teaches Amy what safety feels like.

Every pause builds choice.
With patience and practice, children learn that they can feel big things and still stay in control.

Resources & GROWTH

Resources:

(Amygdala + fear response, emotional regulation)

- Society for Neuroscience (BrainFacts). The neuroscience of stress (amygdala and stress response).
- NIMH. Guided visualization: dealing with stress (amygdala/fight-flight-freeze explainer).
- NINDS (NIH). Brain Basics: Know Your Brain / Brain Basics resources.
- Cleveland Clinic. Brain: parts and function (overview pages).
- CASEL. CASEL's SEL Framework (self-awareness, self-management, social awareness, relationship skills, responsible decision-making).
- CDC. Child Development: Positive Parenting Tips and developmental guidance by age.
- MedlinePlus (U.S. National Library of Medicine) (fear response)
- Encyclopedia Britannica (amygdala)

Growth:

(If you or someone you know wants to understand big feelings or calming strategies)

- Child Mind Institute
  - Search: anxiety, emotional regulation
- CASEL
  - Look for: emotional awareness

Acknowledgments:

To my students: past, present, and future:
You taught me more than any textbook ever could. By watching how you explored, questioned, struggled, and discovered, I learned how minds truly grow. Every "aha" moment in these pages began with you.

To the parents, teachers, and guides walking beside children every day:
Thank you for listening, for advocating when things feel complicated, and for believing that understanding a child's inner world truly matters. Your patience and care shape futures in ways you may never fully see.

To my husband, my best friend, and my constant partner in learning:
Thank you for walking beside me, believing in this work, and holding space for creativity to exist. Sharing curiosity, growth, and creation with you has been one of the greatest honors of my life.

To the great thinkers who shaped how I see the world: including Dr. Seuss, Albert Einstein, Leonardo da Vinci, Stephen Hawking, Ludwig van Beethoven, just to name a few, and my English teacher, Mrs. Godfrey:
Your curiosity, courage, and imagination reminded me that learning is limitless, creativity is powerful, and questions matter. Mrs. Godfrey, your encouragement, belief in me, and love of words sparked my writing journey. If not for you, I may never have picked up a pen.

And to my publisher, NextGen Learners™:
It was by fate that we met, and by inspiration that we now move forward. Together, we build something meaningful. Together, we leave this legacy.

Lastly, to my own brain:
Complex, resilient, ever-adapting, and endlessly learning. Thank you for carrying stories, holding wonder, and reminding me that growth is always possible.

Maisel McLaula is the author of the Think Like a Brain!™ children's book collection, a story-first series that blends neuroscience, imagination, and emotional safety for young readers.

Drawing from decades of experience supporting children and families, her work centers on a simple but powerful belief: children deserve to understand and accept who they are, including how their brains work and why they make the choices they do.

Through playful characters and gentle metaphors, Maisel's stories help children discover the "why" behind their thoughts, feelings, and actions, turning confusion into clarity and self-judgment into self-compassion. By understanding their inner systems, children gain the power to respond with kindness toward themselves and to recognize those same patterns in others.

Maisel believes that when children feel safe inside their own minds, empathy grows naturally. The Think Like a Brain!™ collection offers a foundation for confidence, understanding, and care that grows with the child.

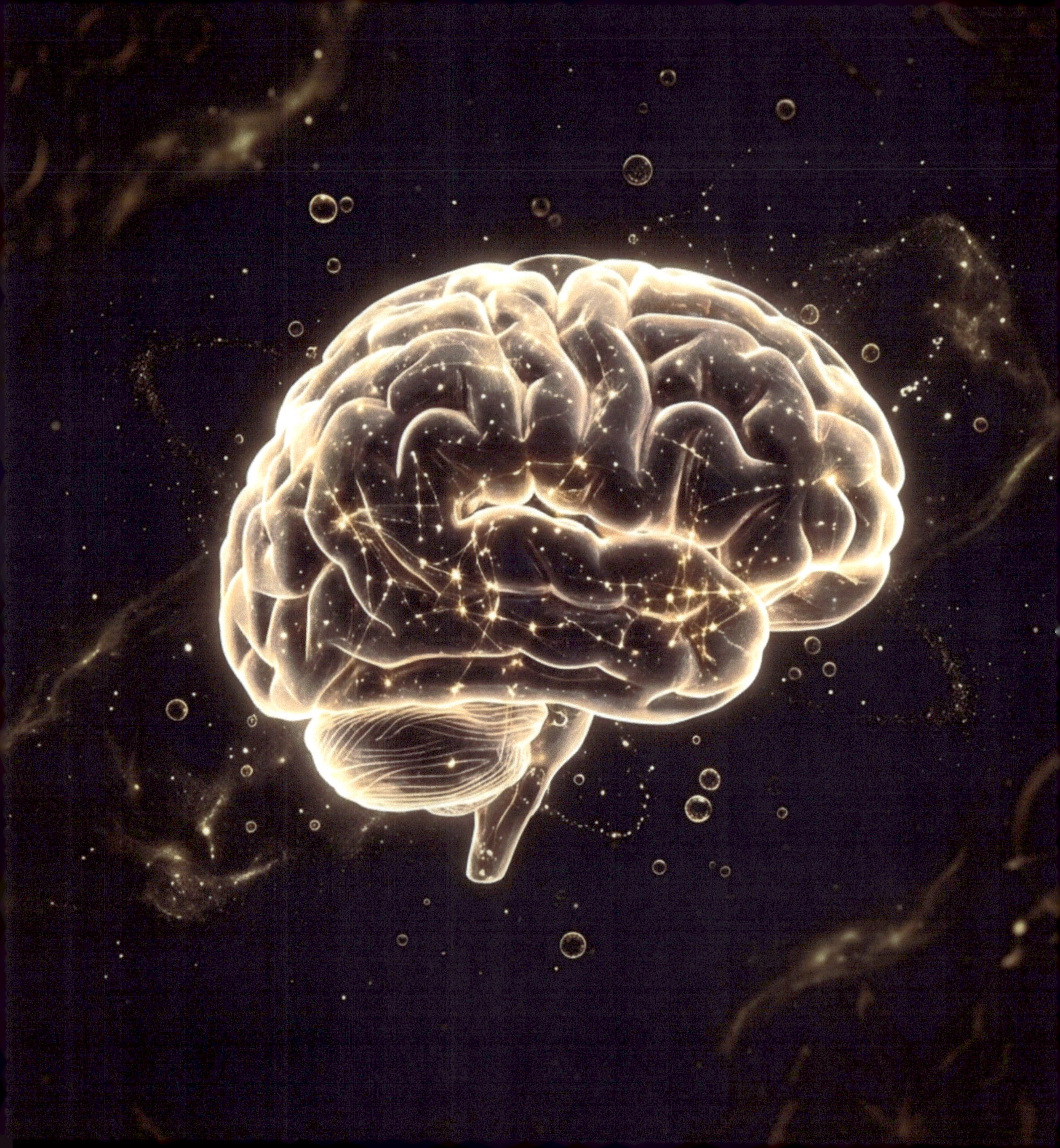

ABOUT THE PUBLISHER

NextGen Learners LLC is an educational publishing company dedicated to helping children understand how they think, feel, learn, and grow.

Through imaginative storytelling, neuroscience-informed concepts, and real-world application, NextGen Learners™ creates books and resources that support emotional development, critical thinking, self-awareness, and confidence in young minds.

Our work is designed not only for children, but for the parents, educators, caregivers, and professionals who walk beside them. Each book serves as a bridge, helping adults better understand the child in front of them, while helping children better understand themselves.

By creating shared language and meaningful connection points, our stories foster deeper relationships, reduce misunderstanding, and encourage acceptance, both within oneself and toward others.

We believe every child's mind is uniquely wired, and that understanding those differences is the foundation for compassion, growth, and lifelong learning. With multiple collections designed to support cognitive, emotional, and developmental growth, NextGen Learners provides tools that children can carry with them, and tools that adults can use to guide them with clarity, patience, and purpose.

NG
NEXTGEN
LEARNERS
PRESS

MORE TO EXPLORE

The Brainy Bunch is just the beginning.
More stories, tools, and brain-friendly resources are on the way! Designed to help children, parents, and educators better understand how the brain works in everyday life.

FREE PRINTABLES

We've created printable pages to go along with this book, including:

-Reading Guides
-Brain helper visuals
-Character Cards

These are available for free as part of our growing resource library.

Our full website experience is built to include:

-Downloadable activity packs
-Parent & teacher guides
-Expanded learning tools
-Additional collections and topics

STAY CONNECTED

We love hearing how children connect with these stories!

If there is a topic, feeling, or challenge you would like to see explored in future books, we would love to hear from you!

Contact us: Inspire@NextGenLearners.com
NextGenLearners.com

Your ideas help shape future stories and support the growth of this collection.

This collection grows with you. New resources and stories are added regularly.

Maisel McLaula loves hearing from the tiny humans who read her books.

Did a story make you giggle, spark a big idea, or inspire you to draw your very own character? She would be delighted to see it! Send your letters and artwork to:

NextGen Learners
Attn: Maisel McLaula
8666 Beechmont Ave #1063
Cincinnati, OH 45255
Every letter is read with a smile, and your artwork just might make her whole day.

OPEN BRAIN

www.ingramcontent.com/pod-product-compliance
Lightning Source LLC
Chambersburg PA
CBHW041322210926
51847CB00002B/22
9781972860076